FENNEC FOX or ARCTIC FOX

BY MARILYN EASTON

Children's Press
An imprint of Scholastic Inc.

A special thank you to the team at the Cincinnati Zoo & Botanical Garden for their expert consultation.

FENNEC FOX

Library of Congress Cataloging-in-Publication Data

Names: Easton, Marilyn, author.

Title: Hot and cold animals. Fennec fox or Arctic fox / by Marilyn Easton.

Other titles: Fennec fox or Arctic fox

Description: First edition. | New York : Children's Press, an imprint of Scholastic Inc., 2022. | Series: Hot and cold animals | Includes index. | Audience: Ages 5–7. | Audience: Grades K–1. | Summary: "NEW series. Nonfiction, full-color photos and short blocks of text to entertain and explain and how some animals with the same name can survive in very different environments"—Provided by publisher.

Identifiers: LCCN 2021044790 (print) | LCCN 2021044791 (ebook) | ISBN 9781338799392 (library binding) | ISBN 9781338799408 (paperback) | ISBN 9781338799415 (ebk)

Subjects: LCSH: Fennec—Juvenile literature. | Arctic fox—Juvenile literature. | Foxes—Juvenile literature. | Habitat (Ecology)—Juvenile literature. | BISAC: JUVENILE NONFICTION / Animals / Foxes | JUVENILE NONFICTION / Animals / General

Classification: LCC QL737.C22 E178 2022 (print) | LCC QL737.C22 (ebook) | DDC 599.776—dc23

LC record available at https://lccn.loc.gov/2021044790

LC ebook record available at https://lccn.loc.gov/2021044791

Copyright © 2022 by Scholastic Inc.

All rights reserved. Published by Children's Press, an imprint of Scholastic Inc., *Publishers since 1920*. SCHOLASTIC, CHILDREN'S PRESS, and associated logos are trademarks and/or registered trademarks of Scholastic Inc.

The publisher does not have any control over and does not assume any responsibility for author or third-party websites or their content.

No part of this publication may be reproduced, stored in a retrieval system, or transmitted in any form or by any means, electronic, mechanical, photocopying, recording, or otherwise, without written permission of the publisher. For information regarding permission write to Scholastic Inc., Attention: Permissions Department, 557 Broadway, New York, NY 10012.

10 9 8 7 6 5 4 3 2   24 25 26

Printed in the U.S.A. 40

First edition, 2022

Book design by Kay Petronio

ARCTIC FOX

Photos ©: cover left: David & Micha Sheldon/F1online/age fotostock; cover right: Matthias Breiter/Minden Pictures; 1 left: David & Micha Sheldon/F1online/age fotostock; 1 right: Matthias Breiter/Minden Pictures; 4: Alain Dragesco-Joffe/Minden Pictures; 8–9: Alexey Sodov/Dreamstime; 10 main: Bruno D'Amicis/Nature Picture Library; 12–13: Bruno D'Amicis/Nature Picture Library; 14–15: Juergen & Christine Sohns/Minden Pictures; 16 center: D. u. M. Sheldon/blickwinkel/Alamy Images; 16 bottom right: TKJim McMahon/Mapman ©; 17 top: Jim McMahon/Mapman ©; 17 bottom: Dejavu Designs/Dreamstime; 18–19: HomoCosmicos/Getty Images; 20–21: Ignacio Yufera/Minden Pictures; 22 main: Bruno D'Amicis/Alamy Images; 23: Sergey Gorshkov/Minden Pictures; 24–25: Havranka/Dreamstime; 26–27: Aterra/Getty Images; 28 main: Tierfotoagentur/Alamy Images; 29: Klein & Hubert/Minden Pictures; 30 left: David & Micha Sheldon/F1online/age fotostock; 30 right: Matthias Breiter/Minden Pictures. All other photos © Shutterstock.

CONTENTS

MEET THE FOXES

Fennec foxes and Arctic foxes are very different types of foxes. Most fennec foxes live in the hot desert in North Africa. They search for food at night, when it is cooler. They eat bugs, fruit, insects, and more.

FENNEC FOX

A fennec fox's big ears help it cool off. **FACT**

Arctic foxes live in the chilly **Arctic**. They hunt for food during the day. They eat **rodents** called lemmings.

ARCTIC
FOX

FACT An Arctic fox's fur color changes all year long.

HUGE EARS

Big ears help them hear prey underground.

COOL FEET

The fur on their feet keeps them from burning.

6

HIDE-AND-SEEK

Their tan fur blends in with the sand.

FENNEC FOX CLOSE-UP

A fennec fox can weigh 2 to 3 pounds (1 to 1.4 kg).

It has tan fur with a black tip on its tail.

STRONG LEGS

Powerful legs help the fennec fox jump 2 feet (0.6 meters) high!

FACT Fennec foxes are the smallest type of fox.

ARCTIC FOX CLOSE-UP

STAY WARM

In winter, the Arctic fox grows thicker fur.

An Arctic fox can weigh 7 to 17 pounds (3 to 8 kg).

During winter, most Arctic foxes are white. Some are blue-gray.

COZY TAIL

It wraps its tail around its body to keep warm.

ADORABLE EARS

Rounded tips on their ears prevent freezing.

FURRY FEET

Fur on their feet grips the slippery ice.

FACT In summer, Arctic foxes have gray or brown fur.

HOT AND COLD FUR

FENNEC FOXES

The fur of both the fennec fox and the Arctic fox matches the environment. It allows them to blend into their surroundings, or to **camouflage**.

FACT

A fox tail is sometimes called a "brush."

The fennec fox's fur color **reflects** the sun's rays and helps to keep it cool. It also has an undercoat that protects its skin from the hot sun. The Arctic fox's thick fur keeps it warm in winter.

ARCTIC FOX

Fennec foxes are also called desert foxes. **FACT**

HOT HOME

Fennec foxes mostly live in the Sahara Desert in North Africa. It is one of the hottest places on Earth. The temperature can reach over 120°F (49°C). During the day, it is too hot for fennec foxes to be out in the sun. They hide in their **dens** and come out during the night. They are **nocturnal**.

BRRRRR, IT'S COLD

Arctic foxes live in one of the coldest places on Earth. The Arctic can get as cold as −58°F (−50°C). Arctic foxes also live inside dens, which are holes in the ground that protect them from extreme weather like snowstorms. Over time, different foxes use these dens.

FACT Some Arctic fox dens are around 300 years old.

DEN, SWEET DEN

Fennec foxes and Arctic foxes live in extreme **habitats**. Fennec foxes live in one of the hottest places in the world. Arctic foxes live in one of the coldest.

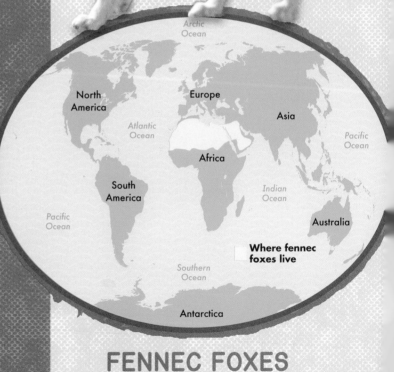

FENNEC FOXES

Arctic Ocean

North America

Europe

Asia

Atlantic Ocean

Pacific Ocean

Africa

South America

Indian Ocean

Pacific Ocean

Australia

Southern Ocean

Antarctica

☐ **Where fennec foxes live**

Both types of foxes live in dens. The dens protect them from the severe environments they live in.

ARCTIC FOXES

Where Arctic foxes live

RUSSIA

FINLAND

SWEDEN

North +Pole

ARCTIC OCEAN (frozen sea)

Greenland (DENMARK)

NORWAY

Alaska (U.S.)

ICELAND

ARCTIC CIRCLE

CANADA

ATLANTIC OCEAN

PACIFIC OCEAN

FACT

Fox dens are also called **burrows**.

FENNEC
FOX

Hyenas are one of the
fennec foxes' **predators.**

NIGHT SNACK

In the Sahara Desert, it is too hot to hunt during the day. That is why fennec foxes mostly eat at night. Fennec foxes are **omnivores**. Fennec foxes eat eggs, birds, reptiles, rodents, insects, plants, seeds, and fruits. There is very little water where they live. Most of their water comes from food.

YUMMY LEMMINGS

Arctic foxes are also omnivores. Arctic foxes eat eggs, berries, birds, and lemmings. They also eat leftover meat from a polar bear's meal. In the winter, when food is hard to find, they will bury their food in the snow or in their den.

ARCTIC
FOX

LEMMING

FACT

Wolves, polar bears, and red foxes are some of the Arctic foxes' predators.

21

UNDERGROUND FOOD

FENNEC FOX

Both types of foxes rely on their excellent sense of hearing to find prey under the ground. Fennec foxes can hear their next meal moving under the sand. Then they dig to catch it.

An Arctic fox can hear a lemming move under the snow. They jump and dive into the snow to catch the lemming.

ARCTIC FOX

FACT

The special jumping the Arctic fox does is called **mousing**.

Fennec fox babies are born in the spring.

FACT

BIG-EARED BABIES

Fennec fox mothers give birth to two to five babies each year. The babies are born with fur. At one week old, they can see. At two weeks old, they learn to walk. They live with both of their parents. They become adults around nine to eleven months.

FURRY BABIES

A mother Arctic fox can give birth to as many as 14 babies. The mother hunts for food while the father watches them. The babies learn from their parents how to be a fox. After about six months, they will be able to take care of themselves.

FACT

A newborn Arctic fox is as small as a kitten.

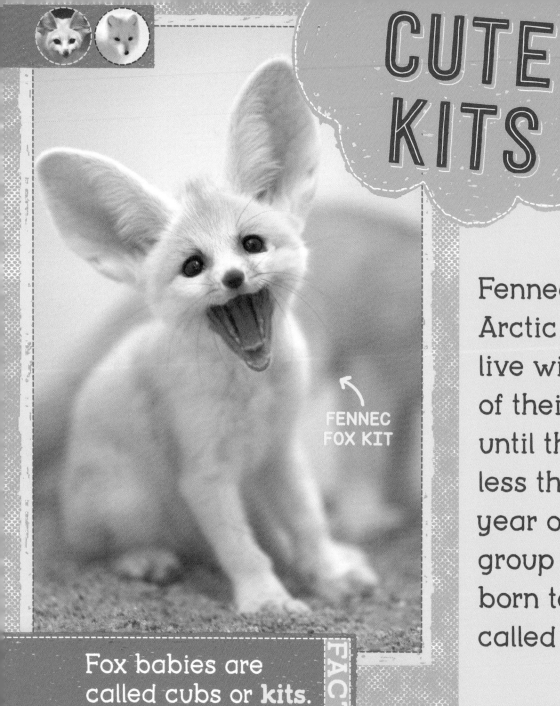

CUTE KITS

FENNEC
FOX KIT

Fennec and Arctic fox kits live with both of their parents until they are less than one year old. The group of kits born together is called a **litter**.

Fox babies are called cubs or **kits**.

FACT

There are many more Arctic fox kits born in a litter than fennec fox kits. This may be because fewer babies are able to survive the tough winter.

ARCTIC FOX KIT

FACT Arctic kits are born with brown fur.

YOU DECIDE!

If you had to pick, would you rather be a fennec fox or an Arctic fox? If you like playing in the sand and digging, you may enjoy being a fennec fox. If you like jumping and spending time in the snow, maybe you would choose to be an Arctic fox!

There are 23 different types of foxes.

GLOSSARY

Arctic (AHRK-tik) – the area around the North Pole

burrow (BUR-oh) – a tunnel or hole in the ground made or used as a home by an animal

camouflage (KAM-uh-flahzh) – to disguise something so that it blends in with its surroundings

den (den) – the home of a wild animal

habitat (HAB-i-tat) – the place where an animal or a plant is usually found

kit (kit) – a baby fox

litter (LIT-ur) – a number of baby animals that are born at the same time to the same mother

mousing (MOUS-ing) – a special pouncing jump an animal does to catch prey

nocturnal (nahk-TUR-nuhl) – active at night

omnivore (AHM-nuh-vor) – an animal that eats both plants and meat

predator (PRED-uh-tur) – an animal that lives by hunting other animals for food

prey (pray) – an animal that is hunted by another animal for food

reflect (ri-FLEKT) – to throw back heat or light from a surface

rodent (ROH-duhnt) – a mammal with large, sharp front teeth that are used for gnawing things

INDEX

ABOUT THE AUTHOR

Marilyn Easton is the author of more than 50 books. She lives in Los Angeles, California, with her two dogs and one cat. If she had to choose, she would be a fennec fox.